EXPLORING
CAREERS

Careers in Travel and Hospitality

ReferencePoint Press®

Other titles in the *Exploring Careers* series include:

Careers in Food and Agriculture
Careers in the Legal Profession
Careers in Renewable Energy

Careers in Travel and Hospitality

Stuart A. Kallen

© 2018 ReferencePoint Press, Inc.
Printed in the United States

For more information, contact:
ReferencePoint Press, Inc.
PO Box 27779
San Diego, CA 92198
www.ReferencePointPress.com

Picture credits:
19: Shutterstock/wavebreakmedia
27: Thinkstock/Dick Luria
49: Shutterstock/CandyBox Images
58: iStock/diverroy
65: Shutterstock/Page Light Studios

LIBRARY OF CONGRESS CATALOGING-IN-PUBLICATION DATA

Name: Kallen, Stuart A., 1955– author.
Title: Careers in Travel and Hospitality/by Stuart A. Kallen.
Description: San Diego, CA: ReferencePoint Press, Inc., 2018. | Series: Exploring Careers | Includes bibliographical references and index. | Audience: Grade 9 to 12.
Identifiers: LCCN 2017034740 (print) | LCCN 2017038664 (ebook) | ISBN 9781682823187 (eBook) | ISBN 9781682823170 (hardback)
Subjects: LCSH: Tourism—Vocational guidance—Juvenile literature. | Hospitality industry—Vocational guidance—Juvenile literature.
Classification: LCC G155.5 (ebook) | LCC G155.5 .K35 2018 (print) | DDC 910.23—dc23
LC record available at https://lccn.loc.gov/2017034740

Contents

Attractive and Accessible Jobs

A growing number of people around the globe are participating in what is known as the experience economy. This economic trend is based on the idea that people are willing to spend more money to take part in events that deliver memories that last a lifetime. Experiences include adventure travel, food tourism, and luxury cruises. Richard Fain, chief executive officer of Royal Caribbean Cruises, explains the experience economy on the Skift website. According to Fain, people are saying "this is the time I can spend with my family, this is the time I can develop memories that will last. And I do think that is somewhat of a cultural shift, and I think [the travel industry is] benefiting from that."

The experience economy is one of several trends driving growth in the travel and hospitality industry. In recent years low airfares and lower-priced gasoline have helped make leisure travel affordable to a broad segment of the American public. And the business travel sector is also booming, according to the consulting company Deloitte. Global business travelers spent a record $1.2 trillion in 2015. According to the US Department of Commerce, all this tourism and traveling creates a lot of jobs. In 2017 one out of every eighteen Americans were employed—directly or indirectly—by the travel and hospitality industry. This translates to 15.8 million people working for hotels, restaurants, bars, casinos, cruise lines, ski resorts, museums, travel agencies, airlines, and other travel- and hospitality-related businesses.

The Happiness Business

As the name implies, hospitality professionals need to be hospitable (or friendly), patient, resourceful, and generous. Customers often expect everything to be perfect, and hotel and restaurant managers, bartenders, cruise ship directors, and flight attendants need to react to complaints with politeness and a positive attitude. As Boulder, Colorado, resort manager Michael Bargas explains on the Colorado Mountain College website, "Unlike many organizations, we are in the happiness business, and have the obligation as well as the opportunity to provide unforgettable experiences to our guests. They have made a significant financial commitment in exchange for our services, and we must never forget the essential role that they play in the continued success of our business."

Travel and hospitality workers also perform numerous tasks behind the scenes to keep operations running smoothly and efficiently. Many are skilled in business accounting, human resources, food and beverage management, and facilities management. They are also familiar with the stressful factors associated with their jobs. Travel and hospitality workers often deal with multiple crises in a single day, including equipment breakdowns, accidents, food safety issues, and in some cases criminal activity and even terrorist threats. And the hours can be long. Flight attendants can expect 4:00 a.m. wake-up calls, flight cancellations, and delays. Hotel and restaurant workers might be expected to work twelve-hour shifts, while some begin their careers on the night shift, working from midnight until 8:00 a.m.

Many are attracted to the hospitality industry because of the job perks. Those who work for large hotel chains can receive great benefits, including 401(k) retirement plans, health care packages, and vacation pay. These workers might also have access to free or deeply discounted hotel rooms, pools, ski runs, fitness centers, and food and entertainment venues.

Surveys also show the travel and hospitality industry boasts a higher chance for potential advancement than any other economic sector. Ambitious individuals who start with entry-level positions

Careers in Travel and Hospitality

Occupation	Minimum Educational Requirements	2016 Median Pay (projected)
Bartender	No formal educational credential	$20,800
Chef and head cook	High school diploma or equivalent	$43,180
Flight attendant	High school diploma or equivalent	$48,500
Food and beverage serving and related workers	No formal educational credential	$19,630
Food service manager	High school diploma or equivalent	$50,820
Lodging manager	High school diploma or equivalent	$51,840
Meeting, convention, and event planner	Bachelor's degree	$47,350
Sales manager	Bachelor's degree	$117,960
Security guard and gaming surveillance officer	High school diploma or equivalent	$25,840
Waiter or waitress	No formal educational credential	$19,990

Source: Bureau of Labor Statistics,
Occupational Outlook Handbook, 2015. www.bls.gov.

can expect promotions—servers become head chefs, bellhops become hotel managers, and bartenders become food and beverage managers.

People who love to travel are likely to be attracted to careers in the travel and hospitality industry. As career counselor Angela Rose writes on the HCareers website:

> Hospitality jobs exist in every country of the world. And whether you're a sous chef, room attendant, night auditor or host/hostess, your skills should transfer to any locale. From cruise ship jobs to hotels, restaurants and resorts in other lands, you'll find plenty of opportunities to feed your travel bug, explore different cultures, and meet new people when you work in the hospitality industry.

Attractive and Accessible

Students who want to feed their travel bug can find numerous schools in the United States that offer hospitality and tourism programs. Students can learn basic work skills at vocational schools or earn bachelor's, master's, or PhD degrees in restaurant, hotel, and resort management at universities. Kristin Lamoureux, senior tourism research scholar at George Washington University, says whatever the student's level of education, "hospitality and tourism jobs are attractive and accessible."

Statistics back Lamoureux's claim. According to the World Travel & Tourism Council, the travel and hospitality industry accounted for 8 percent of all jobs throughout the world and generated around $3.5 trillion in 2016. In the United States around 128,000 new jobs were created by the travel and hospitality industry between January and April 2017 alone.

The travel and hospitality sector is growing faster than almost any other industry. Resort managers, cruise ship directors, tour guides, and other travel professionals work in some of the most beautiful places on earth. Every day in the office is filled with new challenges and the exciting sights and sounds most people only experience while on vacation.

Lodging Manager

What Does a Lodging Manager Do?

At a Glance
Lodging Manager

Minimum Educational Requirements

High school diploma; associate's degree preferred

Personal Qualities

Leadership, business skills, customer-service and communication skills, problem solver

Certification and Licensing

Certified Hospitality and Tourism Management Professionals designation from the American Hotel & Lodging Educational Institute (voluntary)

Working Conditions

Long hours, including evenings, weekends, and holidays

Salary Range

Average annual wage of $51,840 in 2016

Number of Jobs

As of 2014, about 48,400

Future Job Outlook

Expected growth of 8 percent through 2024

When Fred Sawyers tells people what he does for a living, they often react with wonder and envy. Sawyers is the manager of the Walt Disney World Swan and Dolphin Hotel in the heart of the Walt Disney World Resort theme park in Orlando, Florida. After thirty years of working at various hotels, Sawyers has one of the best jobs a lodging manager can have. But the work is not simple or easy. He oversees eighteen hundred employees in a 2,265-room hotel complex that features a golf course, spa, several swimming pools, numerous shops, seventeen restaurants and lounges, and 330,000 square feet (30,658 sq. m) of meeting space. But even at a massive resort like the Swan and Dolphin Hotel, the lodging manager has one overriding job, as Sawyers says in *Connect* magazine: "When it comes to enhancing the guest experience, the key is personalizing service." Personalized service involves addressing guests by name, recognizing repeat customers, rewarding loyal lodgers with discounts and gifts, and even sending handwritten apologies if problems arise. Most lodging managers are not responsible for providing personalized service for as many as six thousand guests at a time. But all lodging managers must balance the desires of their guests, the needs of their staff, and the demands made by the company that owns the establishment.

Lodging managers are often referred to as general managers, or GMs. They are the most senior administrators at a hotel or motel. Lodging managers are required to ensure that their establishments are profitable. They set room rates, conduct daily profit-and-loss reviews, and approve budgets and expenditures. Lodging managers monitor guest feedback through surveys and social media reviews and conduct room-to-room quality assurance inspections. And as hotel manager Josh Cushing explains in an interview on the Keep Me Current website, lodging managers must provide strong leadership for employees: "I have tried to foster a positive culture that inspires our team to be the best they can be every day, and this trickles down to our guests. . . . [Strong leaders] develop this culture in their hotels to reach consistent peak performance."

Lodging managers are responsible for operations in what is called the front of the house, or areas open to the public. The front of the house in a hotel or motel includes the front desk, restaurants, banquet rooms, meeting rooms, guest service desks, and decor. Lodging managers also oversee the back of the house operations, which include kitchens, service hallways, laundries, housekeeping and maintenance areas, and refuse disposal.

Lodging managers who work for small hotels and motels oversee the entire operation. Larger facilities usually employ a number of managers who have similar responsibilities as general managers but administer limited aspects of the operation. Resident managers, as the name implies, live in an apartment on the hotel property and are on call twenty-four hours a day. Resident managers oversee staff training and services, prepare and carry out marketing and sales plans, deal with safety and security issues, and ensure customer complaints are taken care of. Revenue managers oversee all monetary matters.

They monitor room sales and reservations, balance budgets, oversee accounting and cash flow, and decide which rooms to offer at discounted rates. Convention service managers coordinate meetings and banquets. They meet with event planners to determine food and room requirements for conventions and are available throughout an event to answer questions and handle problems. Front office managers are responsible for the front desk and oversee guest check-in and check-out operations.

How Do You Become a Lodging Manager?

Education

Traditionally, lodging managers were picked from the ranks of the assistant managers or other hotel workers who started out at entry-level jobs such as bellhop. Those who exhibited strong management skills would rise through the ranks until they were promoted to GM. In recent years the lodging business has become more complex and competitive, and many institutions require managers to hold a two-year associate's degree in hotel management from an accredited technical institute or vocational school.

Most large, full-service hotels expect lodging managers to have at least a bachelor's degree in hospitality management or a related field. A college student aiming for a career in hospitality management can specialize in programs aimed at the operational side of the business or the corporate side, which includes jobs in areas such as finance, marketing, and human resources. Those who possess a bachelor's degree in hospitality management will find more opportunities for advancement. Bachelor's degree programs focus on hospitality sales and marketing, meetings and events, hotel operations, gaming and resort management, tourism development, and restaurant management.

Those who wish to obtain a master's degree in hospitality management study advanced concepts in sales, marketing, accounting, and strategic management. Graduate students can obtain degrees such as a master of management in hospitality, a master of science in hospitality business management, or a master of science in hotel administration. Graduate degrees form a foundation for obtaining

PhD degrees with titles such as business administration in tourism and sports or business administration in hospitality and tourism.

While a large number of colleges and universities offer degrees in travel and hospitality, several are widely recognized for their high-quality programs. The Cornell University School of Hotel Administration, founded in 1922, was the first in the world to offer undergraduate hospitality management degrees. Students who attend the school can study abroad in over forty countries and gain experience working at the luxurious Statler Hotel in Ithaca, New York. The School of Hospitality Business at Michigan State University has the nation's second-oldest hospitality education and research program. The school's Career Expo attracts over eighty hospitality companies every year and provides internship and job opportunities for students. The William F. Harrah College of Hotel Administration at the University of Nevada is located in Las Vegas, which is home to casinos, resorts, hotels, and other tourist destinations. Students in this program have numerous opportunities to study while working at the city's leading restaurants, resorts, and casinos.

Certification and Licensing

Lodging managers are not required to be certified or hold a professional license, but prospective GMs will find that certification will lead to greater employment opportunities and higher salaries. The American Hotel & Lodging Educational Institute (AHLEI) offers industry-recognized certification through its Hospitality and Tourism Management Program (HTMP). The online program teaches lodging management basics such as hotel administration, accounting, hotel maintenance, housekeeping, food service management, and marketing and sales. The HTMP is open to high school students and those who already work in the industry. The program is offered at various institutions in twenty-six states plus the District of Columbia.

Graduates of the HTMP receive accreditation as Certified Hospitality and Tourism Management Professionals. College students and hotel workers can obtain a Certification in Hotel Industry Analytics through the AHLEI. Certification can also be obtained through the Accreditation Commission for Programs in Hospitality Administration, which offers two types of certification, one for those with an associate's degree and another for those with a bachelor's degree.

Internships and Work Experience

Lodging managers are required to interact with the public, think outside of the box, and make decisions under pressure. Most employers in the lodging industry understand that these skills cannot be taught in the classroom—prospective managers need practical experience that can only be gained on the job. For this reason most bachelor's degree programs in hospitality management combine classroom learning with work experience through internships. Joan Tuaño, a graduate of Les Roches International School of Hotel Management, describes her internship at a Ritz-Carlton in Penha Longa, Portugal:

> My first four months were taking the role of Club Level Concierge wherein I performed high levels of personalized service for the crème de la crème of our hotel guests. Then I had to learn how to make a bed and clean bathrooms in housekeeping departments for three months. I also spent some time in the laundry team where I learned scheduling and more managerial tasks. . . . Mastering all aspects of hotel operations specifically in the back of [the] house is critical to succeed in the future in hotels.

Not everyone has the educational credentials to land an internship, but there are other ways to gain experience in the hotel industry. Many hotels and motels provide summer jobs to high school students. While part-timers do not manage staff, they get an insider's view of the hospitality business. A part-time worker can also make valuable contacts that might prove helpful when pursuing a career as a lodging manager in the future.

Skills and Personality

Lodging managers have very public jobs that require strong communication and customer service skills. They greet guests, interact with hotel staff, take numerous phone calls, and must remain polite and calm under pressure, even when dealing with unruly guests. A talent for organization is required since lodging managers often juggle multiple problems throughout the day. Leadership skills are important since lodging managers work with a large staff. Managers must be adept at motivating workers and resolving worker conflicts. Lodging managers also need a good head for business since the job requires attention to budgets and marketing.

On the Job

Employers

Lodging managers work in the traveler accommodations industry, which includes hotels, motels, bed-and-breakfasts, resorts, and unique places like dude ranches and guest camps. About 48,400 lodging managers were employed in 2014 in the United States. About half were self-employed as owners of small hotels, bed-and-breakfasts, and inns.

Working Conditions

Holidays, evenings, and weekends are the busiest times for most lodging managers, and many are on call twenty-four hours a day. This is especially true for managers who live on-site. Despite the work schedule, a 2014 CNNMoney survey showed that most hotel managers give their job a B grade for personal satisfaction and job security. The job also offers certain perks. Live-in managers, for example, have a free apartment, can eat for reduced rates at the establishment's restaurant, and have access to on-site pools, gyms, and spas. Those who work for large hotel chains like the Radisson or Holiday Inn have opportunities to be transferred to hotels located overseas or in exotic locales.

Earnings

According to the Bureau of Labor Statistics (BLS), the average pay for a lodging manager in 2016 was $51,840 annually. This is considered a good wage for a job that does not necessarily require a high level of education; the entry-level requirement for a hotel manager is a high school diploma or one to five years of relevant work experience. However, those with college degrees have a better chance to earn higher wages. The BLS states that the highest-earning group of lodging managers, those in the top 10 percent, bring in more than $96,570 annually.

Opportunities for Advancement

There is good room for growth in the field. Lodging managers often begin their careers as assistant managers and work their way up from positions like resident manager, front desk clerk, or housekeeping executive. Those who work for large hotel chains can advance to executive positions in which they oversee numerous hotels in a state or region.

What Is the Future Outlook for Lodging Managers?

The tourism industry continues to expand, contributing to a mounting need for lodging managers. Employment is expected to grow by 8 percent through 2024, according to the BLS. Much of this growth is expected to take place in the full-service hotel, casino, resort, and convention market, where customers are willing to pay more to experience luxury vacations.

Find Out More

Accreditation Commission for Programs in Hospitality Administration
PO Box 400
Oxford, MD 21654
website: www.acpha-cahm.org

This organization offers two certification programs, the Accreditation Commission for Programs in Hospitality Administration for those with a bachelor's degree and the Commission for Accreditation of Hospitality Management Programs for those with an associate's degree. The website features training videos, webinars, and other relevant information.

American Hotel & Lodging Educational Institute (AHLEI)
6751 Forum Dr., Suite 220
Orlando, FL 32821
website: www.ahlei.org

This organization provides industry-recognized education and training programs to prospective hospitality professionals in high school, in college, and at workforce agencies. The AHLEI works with schools to develop curriculum packages, offers online courses, and provides scholarships. Graduates are accredited as Certified Hospitality and Tourism Management Professionals.

Hospitality Asset Managers Association (HAMA)
website: www.hamagroup.org

HAMA is a network of professional hospitality managers. The association offers a Certified Hotel Asset Manager designation to professionals knowledgeable in all facets of hotel operations. The website has career information, job listings, and educational information.

National Tour Association (NTA)
101 Prosperous Pl., Suite 350
Lexington, KY 40509
website: http://ntaonline.com

The NTA is a business association for travel and hospitality professionals. It offers a newsletter and online magazine that provides up-to-date information about industry trends and growth. The website offers webinars and educational programs beneficial to students interested in the hotel industry.

Food and Beverage Manager

At a Glance
Food and Beverage Manager

Minimum Educational Requirements

High school diploma or equivalent

Personal Qualities

Good communicator, organizational skills, good with people, physically fit

Certification and Licensing

Foodservice Management Professional designation from the National Restaurant Association Educational Foundation (voluntary)

Working Conditions

Long hours, including nights, weekends, and holidays; lots of walking, standing, and lifting

Salary Range

Average annual wage of $50,820 in 2016 (all food service managers)

Number of Jobs

As of 2014, about 320,600

Future Job Outlook

Expected growth of 5 percent (all food service managers) through 2024

What Does a Food and Beverage Manager Do?

When eating a fine meal at a great restaurant, most people do not think about the person who ensured the tomatoes were fresh, the meat was tender, and the wine was sublime. In fact, few are even aware that restaurants and bars employ a person whose job title is food and beverage manager, referred to in the industry as the F&B manager. As the name implies, an F&B manager purchases and tracks an eating establishment's inventory of bread, meat, dairy products, vegetables, fruit, grains, beans, sugar, flour, soft drinks, beer, wine, and spirits. F&B managers are food service professionals who ensure that adequate supplies are always on hand to cook every item on an establishment's menu. They ensure that the products are fresh, are wholesome, and meet government mandated safety standards.

In restaurant lingo, F&B managers oversee operations in both the back of the house and the front of the house. The back of the house refers to behind-the-scenes areas off-limits to customers—most importantly, the kitchen. This is the central command center of any restaurant—where food is prepared, cooked, and plated.

In addition to overseeing the food and drink supply chain, F&B managers often assume other management responsibilities within the fast-paced restaurant environment. Some F&B managers supervise food and drink preparation staff, including chefs, cooks, bartenders, and servers. They act as team leaders who motivate their employees to do their best work, discipline employees, and hold regular staff meetings.

The front of the house refers to all operations that customers visit in a restaurant, including the entry area, waiting area, bar, dining

A restaurant's food and beverage manager talks with the chef about ingredients and other supplies that need to be ordered. Besides overseeing the food and drink supply chain, some F&B managers also supervise staff.

room, outdoor seating area, and restrooms. Customer service is one of the most important front of the house tasks for an F&B manager. When problems arise with a food or drink order, the kitchen runs out of ingredients, or service is slow due to high demand, the F&B manager interacts with diners, handles serving glitches, and addresses whatever problems may arise. Other front of the house duties include overseeing dining room preparation such as the setup of tables, chairs, linens, table settings, and glassware.

F&B managers ensure that servers work efficiently and bartenders adhere to complex regulations that govern alcohol sales. In addition, F&B managers confirm that all service staff are in proper uniform. In some cases they help design staff uniforms to ensure the outfits reflect the look and brand of the restaurant.

F&B managers often work long hours. When restaurants are very busy, F&B managers will perform whatever jobs are necessary to ensure guests are served in a timely manner. They might bus dishes, deliver food to tables, or help out in the kitchen. They also perform various duties before and after restaurants are open to customers.

One of the most challenging jobs concerns analyzing and changing food and drink offerings on the menu. F&B managers consult with chefs and restaurant owners to create dishes that meet an establishment's sales and revenue goals while working to satisfy the ever-changing appetites of customers. This requires an F&B manager to keep up with the latest food fads and trends.

Some F&B managers are specialists who focus on beverages, including beer, wine, spirits, and cocktails. Wine managers are known as sommeliers. Sommeliers usually work in upscale restaurants, where they serve and pour wine and help customers with selections. Sommeliers are responsible for stocking wines, along with specialty beers and spirits. They need to know the best wines that can be paired with each dish in a restaurant. Sommeliers are familiar with the numerous regions where wine is produced in California, France, Germany, and elsewhere, and they understand the subtle differences between varieties.

Michael Shetler is a beverage manager whose job involves mixology, or the art of creating new and unique cocktails. Shetler works at the huge Aria Resort & Casino in Las Vegas, Nevada, where he oversees 350 beverage employees who work in the resort's bars and restaurants. Shetler describes his job in an interview with *Las Vegas Weekly*: "America is very serious about its mixology, so we have to raise the bar. . . . We change our drink menus frequently, do things seasonally and bring people in from outside for seminars and instruction. We offer as much education as we can so employees stay interested."

How Do You Become a Food and Beverage Manager?

Education

A formal education is not necessarily required for those hoping to become F&B managers. However, most employers will only hire F&B managers who have had at least two years of experience in the restaurant industry. High school students interested in gaining food and beverage experience can participate in the ProStart program offered by the National Restaurant Association Educational Foundation. This two-year curriculum combines classroom study with on-the-job training.

Prospective F&B managers who are aiming for higher salaries at upscale restaurants and hotels should consider a postsecondary education. There are no specific degrees for F&B management, but students can work toward obtaining a bachelor of science degree in restaurant and hospitality management, culinary science, or food service systems administration. Four-year programs focus on restaurant operations, management skills, human resources, accounting, and business. The programs combine classroom learning with practical training at university or local restaurants.

Certification and Licensing

Certification as an F&B manager is available to anyone with classroom training or several years of experience in the food service industry. High school students enrolled in the ProStart program can earn a ProStart Certificate of Achievement after finishing the course. The National Restaurant Association Educational Foundation also offers the Foodservice Management Professional designation to managers and supervisors who complete the association's course work and pass an exam.

Internships

Most four- and five-star hotel chains, such as Hyatt and Sofitel, offer F&B management internships, which paid from $11 to $16 an hour in 2017, according to the website Glassdoor. The jobs are generally open to college students pursuing degrees in hospitality. Interns learn about corporate culture, customer service, financial processes, purchasing procedures, health and safety standards, and much more. Oliver Tang was a student at Cornell University in Ithaca, New York, who worked as an intern at two banquet shifts a week at the Statler Hotel located on campus. On the Hospitality Net website, Tang describes how his internship helped advance his career after graduation: "My experience in the banquet department at Statler gave me the full exposure to the back of the house operation of the F&B department. I had worked in front of the house roles before, so this back of the house work completed the full-picture of hotel operations for me."

Skills and Personality

F&B managers need a combination of skills to smoothly interact with customers and staff, manage food and beverage inventories, and advance the commercial interests of their employers. They understand all aspects of restaurant operation and use their business skills to set prices, budget for supplies, and earn profits from food and beverage offerings. A high degree of organization is required to keep track of schedules, inventory, and financial issues. As team leaders, F&B managers need good communication skills to give clear orders to staff, solve conflicts, and deal with suppliers on the phone.

In an era when restaurant reviews on social media websites can make or break a restaurant, F&B managers are on the front lines of customer service. They must be courteous and attentive to patrons and resolve issues under sometimes stressful conditions. F&B managers must also be physically fit. The job sometimes requires them to lift and move heavy beverage cases, vegetable crates, and large slabs of frozen meat that can weigh more than 50 pounds (22.7 kg).

On the Job

Employers

F&B managers find work at hotels, restaurants, casinos, and resorts. Those who understand the latest trends and tastes are finding work in the growing number of restaurants that offer specialties like Japanese ramen, Hawaiian poke, and vegetarian meals. Brewpubs, establishments that brew their own beer to serve with meals, are another developing segment of the industry, in which F&B managers support the work of chefs who create dishes that can be paired with different styles of beer.

Working Conditions

F&B managers work full time and often put in long shifts. The busiest times are nights, weekends, and holidays. Roger Vieira has been in the restaurant business for thirty years. He began as a dishwasher, worked as a waiter, and eventually became an F&B manager. As Vieira explains on the HCareers website, "If you put in eight hours, you haven't put in your full day. A 12-hour day is typical. It's busy all the time." And while a twelve-hour day might be typical, some shifts can stretch up to sixteen hours. F&B managers are required to walk and stand for many of these hours. They might also be required to stoop, bend, climb, and kneel when performing their duties, as well as working in an environment where kitchens are hot and humid, floors are slippery, and walk-in refrigerators and freezers are very cold.

Earnings

The Bureau of Labor Statistics (BLS) does not keep specific earnings statistics for F&B managers. However, the BLS says that in 2016 all food service managers, including F&B managers, earned a $50,820 median wage (that is, the wage in the middle of high and low earnings). The lowest-paid food service managers earned around $29,000, while the highest-paid brought in $87,120.

Opportunities for Advancement

Tyson Warren provides an excellent example of the way an F&B manager's career can advance. Warren attended Concord University in

Athens, West Virginia, where he earned a bachelor of science in hotel and food service administration. He was hired as an F&B manager by the Hyatt Regency Orlando International Airport hotel in 1999, and his career took off. He was transferred to the prestigious Hyatt Regency Atlanta and promoted to beverage manager, director of outlets, convention services manager, and director of banquets. Within a decade Warren was promoted to director of food & beverage at the Hyatt in Sacramento, California. In this executive position, he oversees the entire food and beverage service for all Hyatt hotels in the region.

What Is the Future Outlook for Food and Beverage Managers?

Employment for all food service managers is expected to grow slowly, by 5 percent through 2024, according to the BLS. The BLS also states that population growth and a strong economy will contribute to a greater demand for a growing variety of food establishments in the coming years.

Find Out More

Food and Beverage Association of America (FBAA)
111 E. Fourteenth St., Suite 390
New York, NY 10003
website: http://fbassoc.com

The FBAA is a nonprofit organization dedicated to encouraging education, career growth, and improved safety and labor standards for those who work in the food service industry. The association offers scholarships to students in hospitality management programs.

Food & Beverage Magazine
1930 Village Center Circle #3197
Las Vegas, NV 89134
website: www.fb101.com

This online magazine contains a wealth of information for those interested in a career as an F&B manager. Readers can learn about the latest industry trends and current events and read interviews with F&B managers, hotel managers, chefs, grocers, and other food and beverage professionals.

National Restaurant Association
2055 L St. NW, Suite 700
Washington, DC 20036
website: www.restaurant.org

The National Restaurant Association represents the interests of over half a million food service businesses. The association's website features training, education, and career advancement information. Students can learn about the National Restaurant Association Educational Foundation's ProStart program, scholarships, and certification.

Worldwide Internships
website: http://worldwideinternships.org

This organization has a Hospitality Internship Program that offers food and beverage and other culinary internships in the Caribbean, China, Europe, the United States, and elsewhere. Internships are open to university students and recent graduates who are fluent in English and are age eighteen to twenty-seven.

Chef

At a Glance

Chef

Minimum Educational Requirements

High school diploma

Personal Qualities

Creativity, business sense, detail oriented, ability to multitask and work well under pressure

Certification and Licensing

The American Culinary Federation offers sixteen levels of industry-recognized certification (voluntary)

Working Conditions

Long hours in hot kitchens using knives and other dangerous implements; lots of standing and bending

Salary Range

Average annual wage of $43,180 in 2016

Number of Jobs

About 127,500 in 2016

Future Job Outlook

Expected growth of 9 percent through 2024

This is a golden age for those who love to cook. On any given day cooks can turn on the television and watch celebrity chefs chop, bake, and chew. Chefs like Guy Fieri, Jamie Oliver, and Ina Garten have become cooking superstars who instruct fans about new and exciting meal preparation techniques. Fieri was worth $8 million in 2016, and Oliver earned an eye-popping $172 million that same year, but most chefs do not find themselves basking in fame or rolling in money. In fact, most chefs work long hours for little pay. According to a 2016 survey by the Food Network, many chefs average sixty- to eighty-hour workweeks, and they all work on holidays. And 65 percent of the chefs surveyed made less than $75,000 a year. Little wonder 60 percent of the chefs said they wanted their own cooking shows.

Whatever the statistics, some people love cooking and baking enough to make a career out of it.

Chefs prepare, cook, and plate food for the enjoyment of their customers. Depending on their position, some chefs develop recipes and plan menus, while others specialize in making sauces or creating desserts.

And some find fulfilling careers, as chef and restaurateur Grant Achatz explains on the Epicurious website: "It's really rare in this world to get paid for something you love to do, so . . . if you're truly passionate about [cooking], then that's what you should pursue."

Chefs oversee food preparation and direct kitchen staff to ensure that meals are properly prepared, cooked, and plated. Chefs confirm that ingredients are fresh, supplies are ample, kitchen equipment is safe and functional, and work areas are clean. They hire, train, and supervise all kitchen workers. Chefs develop recipes, plan menus, and determine how dishes will be presented. These critical decisions require a chef to have a broad understanding of eating trends, food suppliers, and even the way foods are produced.

Chefs are versed in the use of a wide range of kitchen and cooking equipment, including blenders, food processors, meat slicers, grinders, deep fryers, ranges, ovens, and walk-in coolers. They use extremely sharp knives, and according to the Food Network survey, "many chefs have cut themselves on the job, gone to get stitches and returned to work to finish out the night. Accidents definitely happen: Almost every chef we surveyed has been injured on the job in some way."

Chefs have different job titles, depending on their duties and range of experience. Executive chefs, sometimes called culinary directors, oversee menus and cooking operations in a number of restaurants, either spread out geographically or concentrated in a casino or hotel. Those who run the kitchen in a single establishment are called head chefs. They oversee all operations and coordinate the work of all kitchen personnel. Most restaurants have a second in command known as the sous chef. The sous chef prepares meals and fills in when the head chef is off duty, on vacation, or otherwise absent. Sous chefs supervise various other cooks found in restaurant kitchens, including grill cooks, fry cooks, and chefs who specialize in desserts, salads, or pastries.

How Do You Become a Chef?

Education

Some of the world's most famous award-winning chefs, including Jamie Oliver and Gordon Ramsay, are self-taught—they never attended culinary school. And most restaurant owners say they are not looking for a fancy degree when they interview prospective chefs. Employers prefer workers with kitchen experience who are committed enough to stay for a year or more.

Those without restaurant experience can obtain a two-year associate's degree in culinary arts from a community college or technical school. Students enrolled in associate's degree culinary programs learn essential skills such as cooking techniques, features of international cuisines, knife and kitchen equipment techniques, food safety and sanitation, food presentation, menu development, and beverage and wine studies. Associate's programs include classroom study, but students spend most of their time in the kitchen.

In 2017 there were about six hundred culinary academies in the United States that offered associate's degrees. But the rise of celebrity chefs and foodie culture has sent tuition at culinary academies sky-rocketing. According to the National Center for Education Statistics, the average 2016 tuition at the most popular American culinary arts schools was three times the tuition at a standard four-year public university. For example, tuition at the Institute of Culinary Education in New York City was $34,000 a year, while tuition at the International Culinary Center in New York City was nearly $48,000. By way of comparison, the national average tuition for public universities was $28,000. However, as award-winning chef José Andrés says in an interview with *Travel + Leisure*, his culinary education was helpful: "Cooking should be structured. Culinary school taught me how to be organized and to clean as I go. These are the skills that will help you succeed and create systems to make your kitchen efficient."

Those who hope to become executive chefs, kitchen managers, or restaurant owners can obtain a four-year bachelor's degree in culinary arts. Bachelor's programs focus on advanced skills in cooking, restaurant finance and marketing, sustainable business practices, and restaurant research and development.

Certification and Licensing

Prospective chefs can obtain a diploma or certificate that can be completed in a year or less. An accredited culinary arts certificate can be obtained through online classes, which feature videos, webinars, and instructional materials that teach culinary skills and techniques. Certification can also be obtained through the American Culinary Federation (ACF), which offers sixteen levels of industry-recognized certification. The lowest level, Certified Fundamentals Cook, is for those with little experience in the food service industry. Higher-level courses include Certified Sous Chef, Certified Pastry Culinarian, and Certified Master Chef. The ACF also offers certification for culinary administrators and educators. In order to obtain ACF certification, a chef must complete three thirty-hour courses: one in nutrition, one in food safety and sanitation, and one in supervisory management. Chefs are required to pass one practical exam and one written exam and document prior work history (minimum work experience for certification can range from six months to five years, depending on the level of certification).

Internships, Mentorships, and Apprentice Positions

The ACF and most culinary academies host mentorship programs that pair students with professional chefs. The ACF selects participants through an online application process. Many of the world's greatest chefs learned discipline, leadership, teamwork, cooking, and business skills from mentors and feel that their mentorship was a life-changing experience. Mindy Segal, chef and owner of Mindy's HotChocolate in Chicago, describes her mentorship under Michael Kornick, chef and owner of the MK restaurant: "[Kornick] taught me how to be a businesswoman and a leader and not just a chef. He taught me how to manage and to talk to people, both staff and guests. He taught me that my window of opportunity is as far open or closed as I want it to be. . . . A lot of who I am today is because of him."

Mentorships provide a good opportunity to students who want to learn by observing first and practicing later. While mentorship programs have a slight overlap with internships, interns generally have more skills. As with mentorships, intern programs are offered by culinary academies and the ACF. Culinary school graduates will likely find it easier to get a job if they put in some time as interns. Paul Canales discovered this after finishing his courses at the prestigious Culinary Institute of America. "When you leave a culinary program you are minimally competent," Canales says in *Chowhound* magazine. "Then you have to do internships, find mentors, and be willing to let people teach you all the things you don't know."

Culinary apprenticeships offer yet another way students can learn the trade. Apprenticeships provide a combination of classroom instruction and on-the-job training. Apprentices work under qualified professional chefs and take online or classroom courses during off hours. Those who participate in the two-year or three-year ACF apprenticeship program receive certification upon completion.

Skills and Personality

Successful chefs are creative food artists who improve old recipes, develop new ones, and find unique ways to present dishes with a flair. While creativity is not necessarily required in typical family eateries, those who can generate positive comments on social media and foodie websites will find their services in demand.

Beyond food artistry, chefs need a good business sense to efficiently produce meals that are profitable for the establishment. Serving fresh, perfectly made dishes to hundreds of customers a day also requires a commitment to quality and an attention to detail. And running a kitchen requires chefs to be excellent at multitasking, as the Culinary Cook website states:

> For example, you must ensure that those mushrooms you're frying don't burn, but you also have to fire them at the right time so they're perfect by the time the steak is completed. But not only that, but you have to be reading your chits [orders] ahead of time so you know what's coming down the pipeline. And all this has to be coordinated so the customer gets his dish on time and hot.

On the Job

Employers

Prospective chefs will find a vast array of employment opportunities in the culinary world. Chefs can find work at traditional establishments like hotels, restaurants, casinos, hospitals, and corporate headquarters. They can also freelance, working as self-employed private chefs, food truck operators, or even vendors at fairs and street festivals. Some chefs cook for clients with various illnesses or dietary restrictions. Certified executive chef Chris Smith contracted type 2 diabetes while training at the Culinary Institute of America and went on to make a career as the Diabetic Chef. He teaches cooking classes that focus on eating properly and wrote several cookbooks, including *Cooking with the Diabetic Chef*.

Working Conditions

Chefs are on their feet all day in a fast-paced environment. Most chefs work more than forty hours a week, including early mornings, nights, weekends, and holidays. Conditions can be stressful, and there is a high risk of burnout. Kitchens are hot, the floors are slippery, and dangerous objects such as hot stoves, sharp knives, and meat slicers can cause injury. As a result, chefs have a higher rate of illness and injury than the national average. They are subject to slips, falls, cuts, and burns.

Earnings

According to the Bureau of Labor Statistics (BLS), chefs earned an average of $43,180 annually in 2016. Those who earned the lowest 10 percent of wages brought in less than $23,600, while those who earned the top 10 percent brought in more than $76,280. As might be expected, pay was highest in upscale restaurants and hotels in major cities and at resorts.

Opportunities for Advancement

Restaurants generally have high employee turnover. Employees working at entry-level jobs, such as line cook or salad maker, are often able to advance to head chef if they are dedicated to the work. Some chefs go on to start their own restaurants or catering businesses. A number of chefs have written cookbooks, and a select few go on to get their own cooking shows on cable TV.

What Is the Future Outlook for Chefs?

More Americans are eating out every year, and there is a growing demand for high-quality dining experiences. Additionally, customers are demanding new food experiences and greater convenience, which is contributing to the growth of fast-casual restaurants where customers pay for food before eating. These establishments need a larger number of chefs to function efficiently. These trends mean that employment for chefs is expected to grow 9 percent through 2024, according to the BLS.

Find Out More

American Culinary Federation (ACF)
180 Center Place Way
St. Augustine, FL 32095
www.acfchefs.org

The ACF is dedicated to promoting professionalism and education in the culinary arts. The ACF website offers educational and training resources, career advice, apprenticeship programs, and certification to current and future chefs. The Young Chefs Club is open to ACF members age sixteen to twenty-five.

International Association of Culinary Professionals (IACP)
45 Rockefeller Plaza, Suite 2000
New York, NY 10111
www.iacp.com

Membership in the IACP is open to chefs as well as those who work in culinary education and communication. The organization's website features blogs by food writers, cooking school teachers, and others whose insight into the industry can be valuable to students and prospective chefs.

Research Chefs Association (RCA)
330 N. Wabash Ave., Suite 2000
Chicago, IL 60611
www.culinology.org

The RCA is dedicated to blending the culinary arts with the science of food to provide technical information to the food industry. The association offers certification to culinary researchers and provides programs to students interested in learning advanced food preparation, safety, and production techniques.

World Association of Chef Societies
310 rue de la Tour Centra
Rungis, France
www.worldchefs.org

This organization works to maintain and improve culinary standards on a global scale. The association provides information about qualified culinary schools, standards for quality culinary education, certification, competitions, and job opportunities.

Gaming Dealer

What Does a Gaming Dealer Do?

At a Glance
Gaming Dealer

Minimum Educational Requirements

Six- to eight-week gaming dealer training program

Personal Qualities

Friendly, outgoing, basic math skills, good hand-eye coordination

Certification and Licensing

License from a state or tribal gaming commission

Working Conditions

Fast-paced, variable shifts in loud, smoky casinos; standing for long hours

Salary Range

Average annual wage of $19,290 in 2016, plus significant earnings in tips

Number of Jobs

About 94,570 in 2016

Future Job Outlook

Expected growth of 1 percent for all gaming services workers through 2024

The United States is a big country, and many Americans love to gamble. In 2016 more than 80 million people visited the nearly five hundred casinos that operate in thirty-nine states. The casinos are run by hotels, gaming companies, and various Native American tribes. Slot machines generate around two-thirds of casino profits. But millions of gamblers prefer table games such as blackjack, craps, roulette, and baccarat.

The action at table games is overseen by gaming dealers, sometimes called casino dealers or, simply, dealers. Gaming dealers work with gaming equipment like roulette wheels, dice, and playing cards. They deal cards, announce the actions of individual players to the rest of the gamblers at the table, and correct players who make mistakes. A big part of the job is handling gaming chips; gaming dealers collect bets from players and pay off winners. They manage the pot, or money, wagered by other players in a single hand or game. This requires the dealer to verify the amount of the bet, announce the winner, and award the winnings. Other duties of the gaming dealer include calling a floor manager to resolve disputes, requesting employees called chip runners to bring more gaming chips, and alerting waitstaff that a player wants a beverage or food.

Many people want to become gaming dealers because the job involves playing casino games all day as a representative of a casino. Since the odds are stacked in the casino's favor, gaming dealers win more often than they lose. As gaming authority David Sheldon puts it on Casino.org, "If you love gambling, being a casino dealer may seem like the greatest job in the world. You get to play the same games you already love, only from the other side of the table—meaning you'll get to win a little more often than you already do. Sure, you don't get to keep the winnings, but [the casinos] pay you."

There are other positive aspects to working as a gaming dealer. Major hotels and gaming corporations offer dealers good benefits, including health care plans, retirement programs, and two to four weeks of paid vacation every year. Casinos provide free uniforms and free or discounted meals. And since casinos are open twenty-four hours a day, 365 days a year, gaming dealers can work flexible schedules. Some work late nights, early morning hours, or split shifts to deal with personal matters.

While working in a lavish casino may seem glamorous, there are negative aspects to the job. First and foremost is the low pay. While casinos rake in big money at gaming tables, most dealers work for minimum wage. Those with years of experience might earn more, but all gaming dealers rely on tips from players to augment their salary. Gamblers believe it is common courtesy—and that it brings good luck—to tip the dealer. Tips might only come to a few dollars a night or might add up to fifty dollars an hour, depending on the game and the time of day. The biggest tips generally go to poker dealers, but these individuals work hard for their money. Dealing poker requires a high degree of skill: the game is fast paced, and dealers must keep track of numerous pots and players.

Like servers and others who work for tips, gaming dealers are motivated to be friendly and supportive to customers. This means gaming dealers need to be sociable and outgoing. Craps dealer David

Vatthanavong explains in the *Schenectady (NY) Daily Gazette*, "As a dealer . . . you are showing the customer a good time." Gaming dealers engage bettors by providing friendly betting advice, showing concern over losing hands, and offering hearty congratulations when a player wins. As a former casino floor manager known only as Sophie says on the Blackjack Apprenticeship website, "We would always say to dealers, 'You are actors and actresses. Put on a show.'"

Putting on a show while separating gamblers from their money can be emotionally stressful for dealers. Losers sometimes become volatile and blame dealers for their losses. While casinos have good security teams to manage problem customers, dealers can still be subjected to abusive language and threats. Gaming dealers also have to accept the fact that their employers do not trust them. To prevent stealing or cheating, casinos use video cameras, floor managers, and security personnel to keep dealers under constant surveillance.

How Do You Become a Gaming Dealer?

Education

Those who wish to make a career out of dealing at a casino do not need to invest a lot of time or money learning the trade. Prospective gaming dealers can take a four- to six-week training program. Classes offered at vocational schools and community colleges cost around $1,000 in 2017. Private schools were charging anywhere from $1,200 to $2,500, depending on the curriculum. New casinos, before they open, sometimes offer free training to qualified candidates. Gaming programs teach the rules and procedures of each game. Classes also instruct dealers about local and state gaming regulations.

Certification and Licensing

The gambling industry is tightly regulated by local and state laws meant to prevent customer scams, embezzlement, money laundering, and other illegal activities. Native American–run casinos are overseen by tribal gaming commissions. Because gaming dealers handle so much money—and have numerous opportunities to steal—they are required to obtain a license from a state or tribal gaming commission.

A gaming dealer must undergo an extensive background check. Gaming commissioners check an applicant's bank records and character references. They also look for criminal records; anyone with felony convictions or links to criminal figures is automatically disqualified.

As casino operations continue to expand, the lack of qualified candidates with clean records has become a problem. MGM Grand needed three thousand employees for a new casino scheduled to open in 2018 in Springfield, Massachusetts. But according to the *Springfield Valley Advocate* newspaper, more than half of the unemployed in Springfield—around 2,150 people—cannot work at the casino's gaming tables because they have been convicted of crimes such as theft, fraud, perjury, or embezzlement in the past ten years. Anyone with a clean record who has completed a gaming dealer training program would find themselves in great demand at a casino like this one.

Learn by Volunteering

Due to security issues, casinos do not offer internship or volunteer positions to gaming dealers. However, those who wish to gain experience as gaming dealers can volunteer to work for church bingo games or casino night fund-raisers held by charity organizations.

Skills and Personality

Because dealers work for tips, an outgoing personality and the ability to easily interact with customers is one of the most important job skills. "We need someone upbeat, happy, fun and not afraid to try new things, not afraid to meet new people," says Rosemarie Cook, vice president of gaming for Rivers Casino in New York.

Dealers need agile hands to constantly deal cards, work with gaming equipment, and handle chips. Gaming dealers need good math skills. Blackjack dealers have to be able to quickly add up to 21 and multiply chip values that can range from $1 chips to those worth $20,000. Poker dealers also need to be able to split pots evenly among three or more gamblers.

Language skills are a plus, since many gamblers may be visiting from another country and not be fluent in English. For example, a record 207,000 visitors from China traveled to Las Vegas in 2016, and many were drawn to the baccarat tables as well as those offering Asia poker and pai gow poker. At casinos like the Lucky Dragon in Las Vegas, the staff conversed with these customers in Mandarin as soon as they entered the casino.

On the Job

Employers

According to the Bureau of Labor Statistics (BLS), the US casino industry employs 94,570 gaming dealers. They work in commercial and tribal casinos in thirty-nine states where casino gambling is legal. Gaming dealers also work on cruise ships and in casinos throughout the world.

Those seeking work as gaming dealers do not go through a typical job application process. As Sophie points out, gaming dealers are expected to put on a show, and personality plays a large role when casinos make hiring decisions. For this reason, casinos refer to the gaming dealer job interview as an audition. As with an actor seeking a role, job applicants are expected to show their public performance skills as well as their technical abilities to handle cards, chips, and pots.

Working Conditions

Anyone who loves bright lights and constant, fast-paced excitement would make a perfect candidate for a gaming dealer. Casinos are filled with dinging, clanking slot machines and noisy customers who are often inebriated. Casinos are also among the last public places in America where tobacco smoking is still allowed. However, this custom is changing, and many casinos have set aside nonsmoking areas in recent years.

Working as a gaming dealer is physically demanding. Poker dealers work sitting down, but dealers who work other games, including craps tables and roulette wheels, spend their entire shifts on their feet. Casinos compensate for this—and the dizzying work of dealing cards and keeping track of thousands of dollars—by giving gaming dealers a twenty-minute break for every hour of work.

Earnings

Most gaming dealers earn minimum wage and depend on tips to augment their incomes. As a result, the BLS says gaming dealers only earned $19,290 a year in 2016. However, the BLS does not include tips. While the amount of tips a gaming dealer earns can vary wildly, an average dealer makes an estimated $30,000 to $60,000 total a year. While some claim to make over $100,000, this is not considered typical. Overall, the amount a dealer earns is influenced by numerous factors, as Sheldon explains: "When a dealer first starts at a new casino, they may be forced to work at games that are slower-paced, less popular, and generate fewer tips. The demand for dealers can be somewhat seasonal as well, and when the casino doesn't need as many tables open, there won't be as many hours of work to go around, cutting into the earning potential of working in this field."

Opportunities for Advancement

Some gaming dealers begin their careers as part-time or seasonal employees who work during holidays or other busy times. They use this experience to gain full-time employment. Gaming dealers who are honest, dependable, technically skilled, and good with cards and customers can advance their careers. Experienced dealers earn a higher wage, get more desirable shifts, and get to work at higher-stakes tables, where the bets—and the tips—are bigger. Gaming dealers can also get promoted to supervisory or managerial jobs.

What Is the Future Outlook for Gaming Dealers?

The BLS does not keep separate statistics for gaming dealers but projects 1 percent growth for all gaming services workers through 2024. However, this slow-growth figure might not be completely accurate. In 2017 three huge new multibillion-dollar casinos were under construction in Las Vegas, while several existing hotels were expanding their casinos. New casinos were being built in New York, Massachusetts, Mississippi, and elsewhere. New casinos mean more jobs for gaming dealers as the industry continues to expand throughout the United States.

Find Out More

American Gaming Association (AGA)
799 Ninth St. NW, Suite 700
Washington, DC 20001
website: www.americangaming.org

The AGA is a trade group that represents the casino industry. The association's website features news articles, research, and other information of interest to prospective gaming dealers.

California Nations Indian Gaming Association (CNIGA)
2150 River Plaza Dr., Suite 120
Sacramento, CA 95833
website: http://cniga.com

The CNIGA advocates on behalf of Indian tribes and casinos in California. The website provides information about Indian gaming in the state, research, FAQs, and news and events beneficial to those interested in working as gaming dealers at Indian casinos.

Nevada Resort Association
900 S. Pavilion Center Dr., Suite 150
Las Vegas, NV 89144
website: www.nevadaresorts.org

The Nevada Resort Association represents the state's gaming industry. Anyone pursuing a career as a gaming dealer will find useful information in the site's press room, with articles about table games, dealer employment, and other issues.

Unite Here
275 Seventh Ave., 16th Floor
New York, NY 10001
website: http://unitehere.org

Unite Here is a union that represents over one hundred thousand casino workers in the United States and Canada. The union website has information about student partnerships, jobs, and internships.

Cruise Ship Director

What Does a Cruise Ship Director Do?

At a Glance

Cruise Ship Director

Minimum Educational Requirements
High school diploma

Personal Qualities
Outgoing, fun loving, charming, empathetic, good communication and organizational skills

Certification and Licensing
STCW-95 Basic Safety Training certificate (voluntary)

Working Conditions
Free travel to exotic locations, good food, free lodging, long work days, many months away from home

Salary Range
Average annual wage of $52,128 in 2016

Number of Jobs
About 5,000 worldwide in 2016

Future Job Outlook
Expected growth of 4 percent through 2024

In 2016 the cruise ship industry was the fastest-growing sector of the travel business as over 25 million passengers took a cruise. That number is expected to grow to 30 million by 2020, according to projections by the trade group Cruise Lines International Association. The attraction to tourists is obvious. While cruise ships are known for their many and varied food and drink options, they also offer all sorts of entertainment. Cruise ships feature a mind-boggling number of games, physical activities, contests, and even classes in subjects like cooking, scrapbooking, and digital photography. On any given night, a cruise ship passenger can attend a lecture by a famous author, sing karaoke, shake it up at a dance party, see a splashy Broadway musical revue, laugh at jokes told by a celebrity comedian, or listen to poolside bands playing musical styles from jazz to reggae. Cruise ships also offer art auctions, wine tastings, aerial shows, rock concerts, and more. While the entertainment options might seems endless, it is the job of one person—the cruise ship director—to make sure the four thousand or so people on board never get bored.

The cruise ship director is the face of the cruise line and the person who provides information and entertainment options to passengers from the moment of arrival to the final farewell. Cruise ship directors attend every key gathering held on a ship. They serve as emcees and sometimes perform at welcome-aboard receptions, deck parties, and private events. They book all the onboard hospitality and entertainment staff members and organize their schedules. Cruise ship directors manage entertainers, special guest artists, and speakers; put shows together; attend rehearsals; and work with choreographers and lighting, sound, and audiovisual technicians. Job responsibilities also include administrative tasks like budgeting and billing.

Cruise ship directors are also the voice of the cruise line. They use a ship's public address (PA) system to make announcements about entertainment schedules, ports of call, and disembarkation information. In *Travel Weekly*, cruise ship director Troy Linton describes the activities he announces on a typical day using the Carnival Cruise Lines PA system: "There's a golf putting contest at 9:00 a.m., a dance class at 9:30, a trivia challenge at 10:00 a.m., as well as a snorkeling demonstration, an Eat More to Weigh Less seminar and a game of bingo worth $1,100 bucks." The fun continues well into the night, and the cruise ship director stops by every activity to ensure things are running smoothly. The cruise ship director will also visit bars and lounges to speak with passengers.

Cruise ship directors communicate with passengers in another way. Most cruise lines have free newspapers that are delivered each morning under the cabin door. The director writes a column, compiles content, and publishes schedules for events and shore excursions—the port-side day trips that passengers take to local tourist destinations.

Every port of call offers cruisers numerous shore excursions that might include scuba diving, parasailing, or trips to historic sites. Cruise ship directors are expected to know the details about every available shore excursion. Cruisers want to know which trips are best for kids, if lunch will be provided on tours, if car rentals are available, and more. Linton explains his responsibilities: "I feel the shore excursions are a direct reflection on me," he admits. "If I've been on one or I've heard a tour isn't that good I won't recommend it. My job is to make sure everyone enjoys themselves. I have no problem being brutally honest about some of the tours."

The job of cruise ship director involves more than fun and games. Cruise ship directors must be highly organized. Over one thousand people work on a typical cruise ship, and dozens of these employees answer to the cruise ship director, who hires, trains, and manages deputies, assistants, department heads, and others. In addition to solving passenger problems and conflicts, the cruise ship director must deal with difficulties that arise with the staff and crew.

Cruise ship directors are considered officers of the ship, a title that brings numerous responsibilities. Ship officers are required to be familiar with the workings of the 130,000-ton (117,934-metric-ton) cruise ship from stem to stern. They need to have full knowledge of onboard safety and security procedures and constantly update contingency plans in case of emergency. Cruise ship directors keep travel logs, track major events and accidents, and handle employee disputes and injuries.

The job of cruise ship directors can be stressful. They are at sea for months at a time and only get a few hours on shore when a ship is in port. And they work in a relatively confined space with thousands of people where the environment is not always healthy. Passengers (and sometimes crew members) become seasick, and there are occasional outbreaks of gastrointestinal illnesses that can affect hundreds of people. But whatever the pressures, cruise ship directors are expected to smile through it all.

How Do You Become a Cruise Ship Director?

Education

Cruise lines are more impressed with experience than education and tend to promote seasoned entertainers or assistant directors to cruise ship director positions. Prospective cruise ship directors can enhance their employment chances by performing as musicians, comedians, dancers, actors, or emcees. This will help job seekers find employment on cruise ships and provide opportunities to advance to the position of cruise ship director.

Some cruise ship directors transfer from related industries. For example, luxury hotel or restaurant managers with at least five years' experience can make a smooth transition to a cruise ship director position. Others previously found work as tour guides or entertainment directors at resorts.

While a college education is not necessary, a bachelor's degree in hospitality management or a related field can provide an advantage to those applying for the job of cruise ship director. Courses in tourism, business, management, finance, and maritime activities would be helpful to those who wish to understand the technical aspects of the job.

Certification and Licensing

The job of cruise ship director does not require certification or a license. However, safety is as important on a cruise ship as it is on any other ship at sea. Consequently, some cruise ship directors obtain a Standards of Training, Certification & Watchkeeping designation. Known as an STCW-95 Basic Safety Training certificate, the designation is designed to instruct crew members on ways to handle basic shipboard emergencies. The courses provide instruction in firefighting, personal safety, elementary first aid, and personal survival. The cost is around $900, and the courses can be completed by attending various instructional seminars, which can be found online.

On the Cruise Ship Jobs website, travel writer Connie Motz explains the benefits of obtaining an STCW-95 Basic Safety Training certificate in advance of applying for a job as a cruise ship director: "Earning this certificate on your own shows initiative that will surely prove attractive to any potential employer. . . . In today's competitive job market it may pay for you to have an upper hand over other candidates by investing in your own STCW-95 certification. Anything that will help your résumé and cover letter to stand out over another is worth considering."

Internships

Most cruise ship lines offer internships across a wide range of departments. Interns benefit by participating in day-to-day operations and gain hands-on experience that can lead to permanent employment. One example, American Cruise Lines Hospitality Internship Program, offers undergraduate students a chance to live and work aboard a cruise ship for up to sixteen weeks. The program focuses on food service, housekeeping, and customer service to help students strengthen their hospitality skills.

Skills and Personality

Cruise ship directors are in the public eye, and the job is like being onstage at all times. They interact with hundreds of people a day as greeters, cheerleaders, problem solvers, and activity coordinators. Those who work in the profession must be charming, outgoing, fun loving, and empathetic. They need to shake hands, smile, laugh easily, and express empathy. Good communications and organizational skills are a must. Cruise ship directors work in a multicultural environment, and command of a second (or third) language is a plus.

On the Job

Employers

In 2017 approximately two hundred thousand people worked on cruise ships around the world. While there are no exact figures for cruise ship directors, every ship has at least one person in that position, and there are several thousand cruise ships on oceans, lakes, and major rivers around the world. Many of these ships are owned by large corporations, including Royal Caribbean International, Norwegian Cruise Line, Carnival Cruise Line, Disney Cruise Line, and Princess Cruises. There are also several smaller cruise lines.

Working Conditions

Cruise ship directors generally sign six- to eight-month contracts with cruise ship lines. After the contract expires, the director gets six weeks off before beginning another six- to eight-month stint, most likely on a different ship. While many love the adventure associated with this line of work, constant travel can take a toll on family life.

Workweeks on a cruise ship are longer than average, and a cruise ship director can expect to work from forty-five to sixty hours—or more—per week. As Linton states, "This job isn't for everybody. It's—literally—an around-the-clock job. You're never off duty and it takes a special kind of person to enjoy this kind of work. But I love it."

The job of cruise ship director has many perks. They are provided with free food, uniforms, and laundry service when aboard the ship. While most cruise ship workers live two to a cabin—with a bathroom down the hall—cruise ship directors get nice-size cabins to themselves, each with a separate bedroom, office, and bathroom. With no living expenses, cruise ship directors can save nearly all their earnings. Additionally, employees who have been with a cruise line for over a year are offered cruise discounts. They can take their families on a cruise for as little as ten dollars a day per person, which adds up to thousands of dollars in saved vacation expenses.

Earnings

The Bureau of Labor Statistics (BLS) does not keep specific earning statistics for cruise ship directors, but according to the JobMonkey employment website, cruise ship directors earned a median wage of $52,128 in 2016. (The median wage is the wage at which half the workers earn more and half earn less.) According to the same website, the highest-paid cruise ship directors brought in more than $80,000 annually.

Opportunities for Advancement

Some cruise ships are more desirable than others. Cruise ship directors with experience may move from older ships to more prestigious ships that are newer or better equipped. Skilled cruise ship directors can also transfer to ships that visit the most desirable parts of the world. Those who tire of life at sea can use their cruise ship experience to move into other hospitality jobs at major hotels, casinos, or theaters. Some cruise ship directors draw on their entertainment connections to return to the spotlight as actors, comedians, dancers, musicians, or singers.

What Is the Future Outlook for Cruise Ship Directors?

Nearly every cruise ship line was scheduled to expand its fleet between 2017 and 2024. Disney, Viking Sun, Carnival Horizon, MSC Seaview, and others were scheduled to launch bigger ships, each with numerous new entertainment options meant to entice customers. While there are no official government statistics concerning the number of jobs for cruise ship directors, there is little doubt that the industry is growing, and more people will be needed to act as cruise ship directors who run the show from stem to stern.

Find Out More

Carnival Cruise Lines Careers
website: https://jobs.carnival.com

This website for the Miami-based cruise line features job listings on land and at sea, with information about cruise ship careers, the application process, and training videos.

Cruise Lines International Association (CLIA)
1201 F St. NW, Suite 250
Washington DC 20004
website: www.cruising.org

CLIA is a prominent cruise industry trade association dedicated to promotion and education for the cruise community. The Travel Professional section of the CLIA website features career information, webinars, training videos, and links to the career sites of several cruise lines.

Disney Careers
website: https://jobs.disneycareers.com/disney-cruise-line

Disney Cruise Lines hires cruise ship employees through this website, which offers shipboard and shoreside employment. Visitors can scan the site for job openings and contact Disney representatives to discuss available opportunities.

Flight Attendant

What Does a Flight Attendant Do?

At a Glance

Flight Attendant

Minimum Educational Requirements

High school diploma or general equivalency diploma

Personal Qualities

Customer service skills, strong communicator, decisive, physical strength, poise, attention to detail

Certification and Licensing

Flight Attendant Certificate of Demonstrated Proficiency issued by the Federal Aviation Administration

Working Conditions

World travel, serving meals in close quarters, standing and walking for long periods, lifting luggage, pushing carts, turbulence, occasional life-threatening emergencies

Salary Range

Average annual wage of $48,500 in 2016

Number of Jobs

About 97,900 in 2016

Future Job Outlook

Expected growth of 2 percent through 2024

Anyone who has ever been on a commercial airliner is familiar with flight attendants in their crisp uniforms. Flight attendants greet passengers, make sure carry-on bags are safely stowed, demonstrate safety procedures before takeoff, and serve food and drinks during flights. But flight attendants are trained to do much more than serve snacks and help passengers with their seat belts. Airlines are required by law to train flight attendants to keep passengers safe. Flight attendants inspect emergency equipment to ensure it is working. They know how to evacuate airplanes and administer and coordinate emergency medical care. Flight attendants are trained to deal with fires, crashes, terrorist attacks, and drunk and unruly passengers. Long before passengers board a plane, flight attendants are at work. As flight attendant Jackie explains on the Airline Job Finder website, "On international flights, we're there an hour and a half early. The pilots come into the briefing room . . . talk about the flight plan, when it's going to be turbulent, and if there are any security issues. Then the flight attendants check on the emergency equipment, making sure the oxygen bottles are full." After passengers disembark, flight attendants write reports about any safety, medical, or security issues that may need attention.

Anyone who works as a flight attendant will explain that the job has distinct positive and negative qualities. On the positive side, flight attendants travel—a lot. Some work mainly in the United States, while others work international routes. Sometimes their layovers (off-duty hours between flights) are spent in modest hotels in small cities; other times flight attendants have layovers in world-class locales. Even when they are not on duty, flight attendants—and their families—can fly for free or at deeply discounted rates. Aviation blogger Jeanne San Pascual sees another benefit to her job—she meets very interesting people: "From VIPs to athletes and celebrities as well as normal, cool people, there's something fascinating that you can learn from almost everyone you meet during flights."

A flight attendant helps a passenger fit her carry-on luggage in the overhead bin. Flight attendants must make sure that passengers are safe and comfortable, but they also must be prepared to act quickly and rationally in an emergency.

The job of flight attendant has some negative aspects as well. Some attendants work three flights a day, each one carrying several hundred passengers. Attendants are required to smile and say hello to each passenger—often while being ignored. They have to help passengers place heavy carry-on bags in overhead bins, and customers can be very rude when informed that their bags are too heavy or too large to be brought onto the plane. A flight attendant who prefers to remain anonymous tells travel writer Anne McDermott, "[Carry-on bags are] a hot button for a lot of passengers and crew. People abuse the policy, pushing it over the limits, because they don't want to pay the checked-bag fees. . . . We never have enough storage, and the battles begin."

Pay is another issue. Entry-level salary for flight attendants on regional airlines can be as low as $18,000 per year. Many new flight attendants have second jobs. And while flight attendants with seniority can take time off to be tourists, those with less flexible schedules do not have time to take advantage of the travel perks. Flight attendant Jacqueline writes on her blog, *Jacqueline Travel*, that layovers are usually ten hours:

> That kind of kills the idea of being able to see the world. You will certainly see plenty of hotels located close to airports. Remember to take an hour off the layover time for deplaning and traveling to your hotel. Take another hour off your layover time to report to the airport an hour before your flight the next morning. Scratch off another 45 minutes so you can get ready in the morning. That leaves you a little over 7 hours in a new city. You can see a lot in 7 hours, you will just have to skip sleep.

Flight attendants also endure the same tedium at airports as the flying public. They must deal with security screening, overpriced food in airport terminals, and long delays caused by weather and other problems. And just like everyone else, flight attendants can suffer from jet lag, fatigue caused by flying through different time zones. While the work is not for everyone, many flight attendants say they love what they do, as Jaqueline writes: "[I] would not dream of doing anything else. . . . Love isn't blind, I clearly see the faults in flying but for some reason it is what I love to do."

How Do You Become a Flight Attendant?

Education

Flight attendants must be high school graduates or possess a general equivalency diploma. While a college degree is not absolutely necessary, competition for flight attendant jobs is fierce. When Delta Airlines announced it would hire four hundred flight attendants in 2014, fifty thousand people applied within hours. Other airlines experience a similar flood of applicants whenever they announce they are hiring. Applicants with a college degree in hospitality, tourism, public relations, or communications have a better chance of qualifying for an interview with an airline.

One thing applicants do not need to do is pay for flight attendant school. As flight attendant Linda told Airline Job Finder, "Don't waste your time and money going to flight attendant school. They make it sound like you're guaranteed a job after you graduate from their school, but it just isn't true. . . . Flight attendant school never guarantees you a job."

The Application Process

Because of the high demand, most airlines only take applications for a short time—in the Delta case the company only accepted applications for two hours. That means applicants must have their résumés and cover letters ready to send at a moment's notice. Many prepare by reading the popular and respected book *Airborne* by flight attendant Lauren Porter, which provides insight about the job and offers guidance for résumé writing and interview preparation.

Porter says job hopefuls should emphasize their customer service experience and people skills since a flight attendant's primary job is to interact with the public. Applicants should also be sure to highlight whether they are bilingual or multilingual. As Jackie explains, "Depending on what international routes an airline flies, speakers of languages spoken in those cities are needed the most."

Applicants should visit the careers page of airline websites and determine when the companies will be taking applications. Those who are lucky enough to get job interviews will need to demonstrate specific qualities to recruiters. As travel blogger Shannon Ullman writes in the *Huffington Post*, "Pay special attention to your personal appearance. Your dress, demeanor, body language, and personal hygiene will all be critiqued by passengers while you're walking up and down the cabin, and recruiters will be doing the same." Because of the emphasis on personal appearance, airlines will not hire those with facial or multiple piercings, tattoos, or hair that is dyed unnatural colors like green, pink, or blue.

Flight attendant work is physically demanding. Applicants must be in excellent health and be able to pass a thorough medical evaluation. Because of safety concerns, flight attendants must meet certain physical requirements. Height requirements vary, but flight attendants generally need to be able to reach overhead bins (over 5 feet [152 cm] tall) but not be so tall that they hit their heads on the ceiling of the plane (under 6 feet 3 inches [191 cm] tall). Body weight must be in proportion to height, and flight attendants must be able to comfortably fit into the small folding seats in the galley areas of the aircraft. Flight attendants must be at least eighteen years old, hold a valid US passport, and be able to pass a thorough background investigation, which includes drug screening and an FBI fingerprint check. Anyone with felony or some misdemeanor convictions cannot work as a flight attendant.

Certification and Licensing

The Federal Aviation Administration (FAA) requires airlines to train flight attendants for three to six weeks in flight regulations, first aid, and emergency procedures like airplane evacuation. Upon completion of training, the FAA awards flight attendants a Flight Attendant Certificate of Demonstrated Proficiency. Flight attendants are required to update their certificate throughout their careers.

Skills and Personality

Flight attendants on the job interact with hundreds of people in very close quarters. They need excellent customer service skills, including patience, poise, and tact, to deal with travelers who can be irritable, tired, and rude. Good communication skills are a must for flight attendants, who clearly provide directions and safety instructions in person and through the plane's public address system. Flight attendants also need to be decisive and alert, aware of security and safety risks during a flight, and possess the ability to remain calm and make good decisions in an emergency. The job is physically demanding, and flight attendants need to be in good shape. They are required to help passengers with their luggage and spend hours every day walking and standing.

On the Job

Employers

Most flight attendants work for major carriers, including Delta, United, and American. Some work on corporate jets or for chartered flight companies. Almost all flight attendants are members of one of two labor unions: the Association of Flight Attendants-CWA or the Association of Professional Flight Attendants. Long ago, nearly all flight attendants were women. This has changed. According to government statistics, around one-quarter of all flight attendants were men in 2015.

Working Conditions

Airlines operate around the clock, and flight attendants are often required to work nights, weekends, and holidays. However, FAA safety regulations mandate that flight attendants spend only seventy-five to ninety-five hours a month in the air. They typically spend around fifty hours a month on the ground preparing flights, writing reports, and waiting for planes to arrive. However, flight attendants have flexible schedules. Those with several years' seniority can pick and choose the days and nights they wish to fly.

Working in an airplane cabin can be stressful and harmful to the health. Flight attendants experience a higher-than-average number of work-related injuries caused by lifting luggage, turbulence, and exposure to contagious diseases from the general public. Jet lag and extended shifts in pressurized cabins can take a toll, causing irregular sleep patterns, stress, headaches, and other health problems.

Earnings

In 2016 flight attendants earned an average annual wage of $48,500. Those with the least seniority earned less than $26,570, while those who earned the top 10 percent of wages brought in more than $78,650. In addition to their salaries, flight attendants are provided with a payment called a per diem, a stipend meant to cover meals and other expenses incurred during layovers. The per diem is around $3 for every hour spent away from their home base, or $72 for each day. Airlines also pay for hotel rooms for flight attendants. The job comes with retirement benefits and health insurance that also includes coverage for dental and vision care. Many airlines also offer profit-sharing programs.

Opportunities for Advancement

There is a definite hierarchy among flight attendants. Those who are new at the job go through a training period for about a year and are known as junior flight attendants. They are closely watched and receive lower pay and fewer benefits. Eventually, junior flight attendants are promoted to senior status, with commensurate salary increases and more control over work schedules. Those who want to advance further can work toward a supervisory role.

What Is the Future Outlook for Flight Attendants?

Airlines try to keep planes as full as possible, making them slow to expand their fleets and add new routes. This practice means employment for flight attendants is not expected to grow more than 2 percent through 2024. However, a new generation of larger airplanes is coming around 2020, and this might increase the number of flight attendants needed, especially on international routes. Competition for the number of available positions is expected to remain very strong.

Find Out More

Association of Flight Attendants-CWA (AFA-CWA)
501 Third St. NW
Washington, DC 20001
website: www.afacwa.org

The AFA-CWA is a flight attendant labor union that negotiates for better pay, benefits, and working conditions for its fifty thousand members. Students visiting the website can learn about AFA-CWA scholarships and find information about many issues flight attendants face.

Association of Professional Flight Attendants (APFA)
1004 W. Euless Blvd.
Euless, TX 76040
website: www.apfa.org

The APFA is a labor union for flight attendants that publishes the latest industry news, media campaigns, and political actions taken by its members. The website provides a window into the world of flight attendants.

Federal Aviation Administration (FAA)
800 Independence Ave. SW
Washington, DC 20591
website: www.faa.gov

The FAA regulates all airlines in the United States. The website describes the requirements needed to obtain a Flight Attendant Certificate of Demonstrated Proficiency and pertinent information, such as the FAA medical and security requirements and preemployment tests.

Wings Foundation
website: www.wingsfoundation.com

The Wings Foundation is a charitable organization run by American Airlines flight attendants to assist flight attendants in need of financial support due to illness, injury, or disability and those who have sustained damage to their residence through a catastrophic event or natural disaster. Prospective flight attendants can find employment opportunities and gain an understanding of the close-knit community formed by flight attendants.

Scuba Diving Instructor

At a Glance
Scuba Diving Instructor

Minimum Educational Requirements

No degree required but must log 50 to 100 dives and complete 200 hours of training

Personal Qualities

Strong swimmer, excellent physical condition, positive attitude, customer relations skills, good communicator

Certification and Licensing

Diving instructor certification

Working Conditions

Outdoors in underwater environments that can be dangerous, twelve-hour workdays up to seven days a week, lifting heavy equipment

Salary Range

Average annual wage of $53,990 in 2016

Number of Jobs

As of 2016, about 15,000

Future Job Outlook

Expected growth of 37 percent through 2024*

*This is for all commercial divers, which includes diving instructors

What Does a Scuba Diving Instructor Do?

Most people consider scuba diving in crystal blue waters through underwater grottos filled with exotic sea creatures to be part of a dream vacation. For scuba diving instructor Alex Brylske, "[it's] one of those jobs you enjoy doing each and every day of your life." Brylske has been teaching others to scuba dive for over thirty years in dozens of places from the Caribbean to Asia and Australia. Like many others who pursue a career as a scuba diving instructor, Brylske started diving for fun and ended up earning a living teaching the sport to others. Brylske says he remains motivated by the challenges: "What other activity allows a teacher to deal in subjects as diverse as physics, physiology, marine science, mechanics, physical education, psychology and even public relations? You also have to have some pretty good counseling skills. . . . Our job believe it or not is at times tough, demanding and often unappreciated, but it's never boring."

Most scuba diving instructors teach recreational diving to individual clients and groups of tourists. They impart diving knowledge and techniques to students in pools and open water. Most have broad knowledge about marine environments, diving equipment and repair, scuba skills, and safety procedures. Some specialize in night diving, deep diving, or cave diving.

Scuba diving instructors work out of dive shops in resort areas or teach at community recreation centers, schools, or colleges. Those who do not work in the tourist industry might teach scuba diving to rescue workers, biologists, marine researchers, shipyard workers, and underwater construction workers. Some instructors teach diving as part of a rehabilitation program in which they work with wounded veterans or others recovering from injuries.

How Do You Become a Scuba Diving Instructor?

Education

Prospective scuba diving instructors need to learn professional-level underwater skills in what is called a divemaster class. Scuba diver Michael Ange calls divemaster class "the real divers' boot camp." Before students can even qualify to take a divemaster class, they must produce a logbook filled with fifty to one hundred dives, preferably in a wide range of environments.

Divemaster class entails about two hundred hours of training in underwater navigation, dive leadership, emergency procedures, and student support in and out of water. Some academic study is required. The diving physics course covers the physical underwater environment, including water pressure and temperature, light absorption, and other factors. A second important academic subject focuses on diving physiology—how body systems like the muscles, skeleton, lungs, and heart react to a variety of conditions while underwater. Outside the

A scuba diving instructor works with a student on breathing techniques in the safety of a swimming pool before the student ventures into the open ocean. Scuba instructors need to have broad knowledge about marine environments, diving equipment, and safety procedures.

classroom, divemaster training occurs in the water, where students perfect buoyancy control, scuba mask clearing, and procedures to follow when oxygen tanks run low.

Like any other class, divers must pass tests as they learn. They are expected to master the academic subjects and demonstrate diver supervision in water. Tough physical tests are part of the training, and they must be completed within a specific time frame. Divemasters have to complete an 875-yard (800-m) swim wearing mask, fins, and snorkel; a 109-yard (100-m) rescue tow with the diver and victim in full scuba gear; and an extended period treading water.

After completing divemaster training, students take instructor classes called an Instructor Development Course or an Instructor Training Course. These courses provide instruction for teaching diving skills and academics to new divers. Those enrolled in the courses participate in mock training classes in a classroom, a pool, and open water. They practice teaching techniques and learn about the diving business.

There are several organizations that offer divemaster and instructor training, including the Professional Association of Diving Instructors (PADI), the National Association of Underwater Instructors, and Scuba Schools International. There are also smaller training and certification schools. However, the three major institutes are recognized throughout the world, and as Ange writes, "if you're hoping to work in the tropics or bounce around a variety of resort or charter operations, it's best to go with one of the bigger agencies because, by virtue of their size and worldwide distribution, you'll have more job opportunities."

Certification and Licensing

Scuba diving instructors need to be certified almost anywhere they wish to work. PADI, the National Association of Underwater Instructors, and Scuba Schools International offer internationally recognized certificates to those who have successfully completed the divemaster course. To obtain instructor certification, divers take a series of tests and log hours in supervised diving experiences.

The scuba diving instructor certification process includes a two-day evaluation called an Instructor Examination. During this exam divers demonstrate their ability to deliver instructions and safety information, evaluate the skills of other divers, teach and perform rescue operations, and maintain the safety of their students. Candidates are required to create a lesson plan and clearly and concisely present the material in a lecture. Divers must complete a comprehensive exam that covers the diving business, training standards, and diving physiology and physics. Before divers can obtain instructor certification, they must also have current cardiopulmonary resuscitation, first aid, and oxygen provider certifications. These are obtained by international organizations such as the Divers Alert Network or the American Red Cross.

The cost for scuba diver training and certification in 2017 was much less than vocational training in other careers. A divemaster program was around $1,000 (plus whatever diving equipment the student needs). Instructor Development Courses cost around $2,000, while the Instructor Examination was around $500. Students who fail the Instructor Examination and wish to take the exam again have to pay to retake it. Other costs such as instructor manuals, classroom materials, and related travel and lodging expenses can bring the total cost to around $5,000.

Internships

While scuba diving instructors are generally employed in tourism and hospitality, there are numerous internship positions available in medical and environmental research. Qualified college students can participate in the Divers Alert Network Research Internship Program in the areas of dive medicine and research. Interns work with mentors to collect data, develop practical skills, and conduct field, laboratory, and epidemiological studies.

PADI offers marine conservation training internships in which scuba divers work in various locations with mentors who lead teams compiling information on fish species, coral reefs, marine debris, and other matters. Interns receive coaching and professional development skills as they conduct research dives up to five days a week.

Skills and Personality

Scuba diving instructors work closely with all sorts of people, most of whom are on vacation. First-time divers might be very nervous and need reassurance. Some customers arrive with hangovers, while others are overly confident in their diving skills. Whatever the situation, scuba diving instructors need to have excellent customer relations skills. They need to hold a group's attention when speaking, reassure novices, and react professionally when people are rude. As Brylske writes, "People skills are just as important and often more so than diving skills. Patience is perhaps the most important requisite; and a close second is flexibility."

Successful scuba diving instructors need to have a love of nature and a good working knowledge of the incredible life found in the sea, as student divers will often ask their instructors to identify fish, plants, and other sea creatures. Customers will also expect their instructors to have a good working knowledge of various types of diving equipment. However, as PADI instructor Rika explains on the Jupiter Dive Center website, "We can't know everything! What we do know backward and forward is teaching you how to dive, and we can certainly answer all your questions about that."

On the PayScale website Emma Mostrom-Mombelli, a scuba diving instructor in Thailand, lists two other characteristics she believes make an instructor especially marketable: "If there's anything that makes you more wanted in the scuba diving industry it would be language skills and a positive hardworking personality!"

Scuba diving instructors need to be in great physical condition. Individuals who suffer from medical conditions such as asthma, diabetes, heart problems, back problems, or mental health issues cannot obtain diving instructor certification.

On the Job

Employers

According to industry estimates, there are around thirty-two hundred dive stores in North America and the Caribbean, and each one employs two to three people full time and another four people part time. While not all employees are certified scuba diving instructors, most are experienced divers. Resorts employ about twice as many divers in their dive operations. Because the job involves long hours and hard work, there is a high degree of turnover in the industry, and resorts and dive shops are consistently looking for new scuba diving instructors.

Working Conditions

Recreational scuba diving instructors work long days, and the busiest times are during holidays like Christmas and New Year's Day, when tourists flock to tropical resorts. Scuba diving instructors typically start their workdays at 6:00 a.m., getting their gear in order and loading boats with heavy equipment. They might teach two groups of clients in a day, one in the morning and one in the afternoon. After teaching, they have to unload the boat and put equipment away. It might be after 7:00 p.m. by the time the scuba instructor leaves the dive center. If the dive center has a bar, the shop owner often expects the scuba diving instructors to hang around and socialize with clients. Some are also expected to do office work. Scuba diving instructors are paid for each individual client they instruct, which means when they are not working they are not earning money. As Rika explains, "I live and work at a beach resort, but you won't catch me lounging on a beach chair on a work day. . . . We're busy diving, filling tanks, doing paperwork, answering emails, repairing gear, cleaning the shop, and the list goes on. We have fun and love to teach and interact with our guests, but they're on holiday—not us."

Earnings

The Bureau of Labor Statistics (BLS) does not collect statistics specifically on scuba diving instructors but does track information on all divers. According to the BLS, commercial divers earned $53,990 annually in 2016. Scuba diving instructors can expect to earn less when just starting out, around $30,000 a year. Those with extra skills, such as retail experience and a boat captain's license, can expect to earn larger salaries, as they are more valuable to dive centers. The job also has a few perks for those who work at resorts. Scuba diving instructors might be provided with inexpensive accommodations and free or discounted meals.

Opportunities for Advancement

The diving industry provides outstanding part-time employment opportunities, which can turn into full-time employment. Some scuba diving instructors open their own dive shops. And as Brylske says, the job offers good prospects to young people who might not wish to go to college: "I've trained several kids right out of high school who have gone on to have highly successful careers in diving. Some are now working for training organizations, while others are still in the resort industry. A few now own their own businesses. . . . To this day, I've never found another profession that can provide this kind of experience to those who are so young."

What Is the Future Outlook for Scuba Diving Instructors?

According to industry figures, around eight thousand people worldwide become certified scuba diving instructors every year. Fewer than two thousand seek full-time positions, and many others work part time as dive instructors. The BLS predicts that employment opportunities for all commercial divers will expand rapidly, with an expected job growth of 37 percent through 2024.

Find Out More

Divers Alert Network (DAN)
website: www.diversalertnetwork.org

DAN focuses on dive safety, emergency services, health, research, and education. The group provides online education videos, cardiopulmonary resuscitation lessons, medical education webinars, internship programs, and scholarships.

National Association of Underwater Instructors (NAUI)
9030 Camden Field Pkwy.
Riverview, FL 33578
website: www.naui.org

NAUI is a respected diver training and certification organization used by Walt Disney World, the Navy SEALs, the National Aeronautics and Space Administration, and others. The association's website offers detailed information about diver certification programs and various locations where they are offered.

Professional Association of Diving Instructors (PADI)
30151 Tomas
Rancho Santa Margarita, CA 92688
website: www.padi.com

PADI is the world's largest diver training organization. Its website describes a variety of courses, which range from entry level, noncertified recreational diving to advanced, certified master programs. Prospective scuba diving instructors can find everything they need to know about class locations and divemaster certification.

Scuba Schools International (SSI)
1646 SE Tenth Ave.
Deerfield Beach, FL 33441
website: www.divessi.com

The SSI offers training and certification in scuba diving, free diving, and extended range diving in schools all over the world, including Hong Kong, Australia, Spain, and China. The website's Dive Professional section provides information about numerous SSI certifications and how to obtain them.

Tour Guide

What Does a Tour Guide Do?

Tour guide Brandon Presser might have one of the world's greatest travel jobs. He has visited more than one hundred countries and published fifty guidebooks. In 2016 Presser was the host of the Bravo TV show *Tour Group*, which follows a group of tourists around the world—some of whom have never traveled before. Presser describes a country's history and culture and works to dazzle the travelers with fun, exciting, exotic, and just plain weird adventures. On the *ShermansTravel* blog, Presser describes how a tour guide can affect people: "They're going to see the world for the first time, and their preconceived notions about people—and how other people live—are going to be [changed forever]."

Most tour guides are not TV hosts, but their jobs are similar to Presser's. Tour guides are sometimes called tourist guides. They provide entertaining and educational commentary to tourists on the history and culture of various destinations throughout the

At a Glance
Tour Guide

Minimum Educational Requirements
High school diploma

Personal Qualities
Group leader, hard worker, physically fit, good public speaker, patience, bilingual or multilingual

Certification and Licensing
Local licenses or national certification, depending on the location

Working Conditions
Outdoors and indoors at tourist destinations; travel on trains, buses, and airplanes; long hours talking, walking, and standing

Salary Range
Average annual wage of $24,920 in 2016

Number of Jobs
As of 2016, about 38,660

Future Job Outlook
Expected growth of 5 percent through 2024

On a tour of Boston's Fenway Park, home of the Boston Red Sox baseball team, a tour guide regales visitors with facts and stories. Depending on where they work, tour guides may have to be experts on the natural environment, historic buildings, local foods, and more. Some also handle travel logistics.

world. Tour guides are often experts on the natural environment, historic buildings, local foods, and native languages of an area. They are responsible for travel logistics, hotel confirmations, and planning for tour groups. Good tour guides help create unforgettable experiences and lasting memories.

Tour guides work in numerous settings. Some guide customers through traditional tourist destinations such as museums, zoos, and botanical parks. Others conduct themed walks through city streets. For example, ghost tours take visitors to allegedly haunted places in London, New Orleans, New York City, and elsewhere. In Los Angeles tour guides take groups to movie studios and past the homes of celebrities. Other tour guides might focus on culinary adventures, wine tours, or countless other niche excursions that are of interest to the traveling public.

One of the fastest-growing segments of the tourism industry is ecotourism, or sustainable tourism, which has grown by 30 percent since 2006. This sector of the travel business promotes cultural preservation, economic development, and environmentally friendly tourism in developing nations. Ecotourism guides are specialists who have an understanding of sustainability and local environmental issues and the impact of tourism and world trade on local areas. The guides inform environmentally conscious travelers about local customs and history and the region's importance in the natural environment. Ecotourism guides might accompany visitors on long hikes to remote cabins in tropical forests or oversee activities such as shopping at local craft markets or visiting regional music, art, and dance festivals.

Adventure tour guides work in a related field known as eco-trekking, in which visitors participate in activities such as camping, hiking, biking, mountain climbing, or rafting in exotic locations. Adventure tour guides need unique skills. They are expected to hike, bike, backpack, raft, ski, or kayak twenty-plus days a month during the busy seasons. They also have to prep all gear and food before a trip, help guests with their packs and tents, repair gear in the field, and lead camp cleanups. Guides are often expected to provide full meals for their clients, although some have assistants to help cook and serve.

How Do You Become a Tour Guide?

Education

People who work as tour guides come from various educational backgrounds. According to the Bureau of Labor Statistics (BLS), around half of all tour guides held a bachelor's degree in 2016, 22 percent had some college education, and 23 percent had only a high school diploma.

College students who want to pursue a career as a tour guide should consider obtaining a bachelor's degree in history, art, architecture, sociology, ecology, or hospitality. Prospective tour guides can also take courses offered by Professional Tour Management Training, available at thirteen hundred colleges and universities in the United States. The curriculum covers numerous aspects of the travel industry, including terminology, trends, how tours are put together, itinerary planning, public speaking, client welcoming, environmental awareness, city tours, motor coach travel, group psychology, and passenger emergencies. According to Cherie Anderson, founder of the Professional Tour Management Training program, "When I train my students they are fully equipped. They can go and put on their résumés that they know procedures for international and domestic tours, they know safety procedures and requirements for hotels and airlines, and they know tour briefings."

Other training programs include the International Guide Academy, which offers courses in Mexico, Canada, China, the United States, and even on cruise ships. The Miami Dade Community College Travel and Tourism Management Program is one of the most popular in the United States. The Tour Guide Training Corporation of Canada is a leader in local and international tour guide training. The Guild of Registered Tourist Guides offers the Blue Badge Tour Guide Training program throughout England and Scotland. Beyond a formal education, tour guides can enhance their public speaking skills by joining Toastmasters International. This organization helps members improve their communication, public speaking, and leadership skills.

Certification and Licensing

Many cities, including New Orleans, New York City, and Washington, DC, require tour guides to obtain professional licenses. In New York, tour guides must pass a test by correctly answering 97 out of 150 questions. In Washington, DC, tour guides must pass an exam and pass a criminal background check. Many nations, including Austria, Egypt, Italy, Japan, Peru, South Africa, and Sweden, also require tour guides to be licensed.

Tour guides can improve their job chances by becoming certified through the National Association of Interpretation, which represents tour guides who work in museums, zoos, nature centers, historical sites, and other cultural sites. The association offers professional development and training along with the Certified Interpretive Guide accreditation to tour guides it refers to as interpreters. To become a Certified Interpretive Guide, applicants complete a thirty-two-hour course, which includes the history and principles of interpretation and classes that focus on making interpretive programs relevant, organized, and enjoyable. After completing the course, applicants are required to pass a fifty-question test and provide a ten-minute presentation. To qualify for the Certified Interpretive Guide program, tour guides must be at least sixteen years of age and have completed related college course work or logged eight thousand hours in the field (four years of full-time work).

Volunteer Work and Internships

Professional tour guides commonly begin their careers as volunteers. Almost every museum, park, and government-run tourist destination, including the Metropolitan Museum of Art in New York City and the National Park Service, relies on volunteers to keep expenses down. Prospective tour guides can simply pick an institution that appeals to them and check its website for volunteer opportunities. Most of these volunteer positions provide good experience working in and around tourist destinations and with people.

Anderson began her career as a volunteer, working as a children's activities counselor on a cruise ship. She was eventually hired as a cruise tour guide, and she made contacts with tour operators who hired her to escort guests on cruises. That led to a job as a tour director for international tour operators.

Skills and Personality

Tour groups can have anywhere from ten to sixty people, and tour guides are the center of attention, which gives the job a certain showbiz aspect. As group leaders, tour guides are expected to be fun, funny, and engaging while expertly describing the history, art, architecture, geology, and flora and fauna of the area where they are guiding. Tour guides must also get along well with people and care about the needs of everyone in the group.

Good tour guides are excellent public speakers and natural storytellers. They show excitement and enthusiasm about the sights and attractions even if they visit the same places day after day. Tour guides need to speak clearly, calmly, and slowly and project their voices so they can be heard at a distance. Tour guides who are bilingual or multilingual will stand a better chance of getting hired.

Tour guides also need to be good problem solvers and exhibit great organizational and leadership skills. As tour guide Reed Drake comments on the JobMonkey website: "You must be in control and completely organized at all times. You are organizing a group of people, chatting with them, informing them about various places, partying with them, and sometimes even counseling them. You must be a leader."

On the Job

Employers

The Internet offers prospective tour guides an almost endless list of tour agencies, tour operators, and cruise lines around the world. Most websites have a careers link with employer information. Another way to find employers is to visit a local tourism office, which will have numerous free brochures from local tour operators.

Some tour guides work with a destination management company located near a trip destination. Such companies hire local tour guides to assist guests by meeting them at the airport, taking them to attractions, leading city tours, booking restaurant reservations, and assisting with parties and events.

Working Conditions

Most tour guides are people who do not want desk jobs. They usually spend their time outdoors, where they are exposed to the elements, including dust, heat, humidity, cold, rain, and snow. Working with large groups can be stressful; most tours have at least one person who complains, argues, makes bad jokes, or acts rudely to others. The hours can be long; some tour guides work up to twelve hours a day and spend many of those hours speaking, which can lead to laryngitis. Active tours, where people walk long distances, are becoming more popular. "On some tours like in China, it may be miles of walking, which I love," says Anderson. "[It would] really be to your advantage to be in shape."

Earnings

Tour guides must love the adventure and travel more than they love money. In general, tour guides earn from $50 to $150 a day, depending on their location, experience, type of tour, and training. However, many clients tip their tour guides up to $20 a day. Experienced tour directors earn more, between $250 and $350 per day. Destination management company tour guides tend to be highly skilled, and tour managers can earn anywhere from $300 to over $400 per day. There are other benefits to working as a tour guide. Meals, transportation, and accommodations are usually free, and tour guides get to work in places most people only see when they are on vacation.

Opportunities for Advancement

Some people who begin their careers as tour guides work their way up to the position of tour director. While guides usually work in specific locations day after day, tour directors manage group tours. They travel to a destination with a group, ensure that people are booked into hotel rooms, and provide options for transportation, meals, and other necessities. Tour directors can also advance their careers by starting destination management companies.

What Is the Future Outlook for Tour Guides?

While the BLS expects employment for tour guides to grow more slowly than the average, only 5 percent by 2024, some sectors will experience better job growth. Travelers are increasingly seeking new and different experiences, and adventure tour guides will be in greater demand than those who work at more traditional sites.

Find Out More

National Association of Interpretation (NAI)
230 Cherry St., Suite 200
Fort Collins, CO 80521
website: www.interpnet.com

The NAI is a professional association for tour guides who interpret at cultural heritage parks, zoos, museums, nature centers, aquaria, botanical gardens, and historical sites. The group holds an annual career fair, and its website has career development information, job listings, training webinars, and certification programs.

National Federation of Tourist Guide Associations—USA (NFTGA)
888 Seventeenth St. NW, Suite 1000
Washington, DC 20006
website: www.nftga.com

This group represents members of tour guide associations and guilds from across the United States. The NFTGA encourages professionalism, publishes ethics and standards for tourist guides, and hosts an annual conference for its members. Prospective tour guides can learn about the profession by visiting the association's website and downloading NFTGA newsletters.

Professional Association of Wilderness Guide and Instructors (PAWGI)
website: www.pawgi.org

The PAWGI established professional standards for wilderness guides and is a certifying agency for certification programs it developed, including Certified Wilderness Guide and Certified Wilderness Instructor. Those interested in becoming an adventure tour guide can enhance their résumé by completing PAWGI programs.

World Federation of Tourist Guide Associations (WFTGA)
website: www.wftga.org

The WFTGA is a professional organization dedicated to promoting high standards in training and ethics within the tour guide profession throughout the world. The association offers information about tour guiding, licensing requirements in various locations, and training seminars and workshops.

Interview with an Executive Chef

Jordan Gottlieb has been the executive chef, or culinary director, at the Greenbush Brewing Company in Sawyer, Michigan, since August 2015. As an executive chef, Gottlieb oversees all food operations at the Greenbush brewpub and at the company's two other restaurants, the Annex and the Clean Plate Club.

Q: Why did you decide to become a chef?

A: In high school, I worked for an ice cream parlor that also made artisan chocolate products including candies, truffles, and character molds. I enjoyed the artistic aspects of the job so I went to pastry and baking school at Le Cordon Bleu in Pittsburgh with the intent of revolutionizing the chocolate industry! Thinking back, it was really a shot in the dark. . . . I had no previous restaurant experience outside the ice cream parlor and I knew almost nothing about cooking, or food. Fate intervened and I fell in love with the pastry and baking industry and knew that was what I wanted to do.

Q: Can you talk about your education?

A: I went to school in a skyscraper in downtown Pittsburgh with very professional looking classrooms and kitchens. I had six- to eight-week class cycles that consisted of bread production, advanced bread production, cakes and pies, candy and chocolate making, plated desserts, and advanced pastry. All of my chefs were award-winning, accredited chefs with volumes of experience and advanced knowledge of the restaurant industry. Along with kitchen labs every day, we had industry-related classes such as hospitality law, public speaking, cost control, and dining room management all taught by industry professionals. It was great except when I graduated I was left with a massive student loan.

Q: Judging from your experience do you think it is important for a prospective chef to attend culinary school?

A: I have found that whether you attain an associate's degree from a community college or a fancy degree from the most prestigious culinary institution, you are going to end up in a large pool of people that generally have the same knowledge as you. Regardless of the school, those who succeed in the culinary industry have two things in common; they work hard and have a thick skin. An alternative to school is to find a restaurant that you really admire and go to work there washing dishes. Work hard and show interest. It's just about the same as going to school. It will take longer but costs a lot less.

Q: What previous chef jobs did you have before moving into your current position?

A: Upon graduation I had two internships in Chicago; in the pastry department at a popular wine bar and on a baking team at a kosher bakery. My first job in Chicago was working as a head baker for a café. Later I worked as a bakery manager for a vegan bakery and restaurant. Following my move to Kalamazoo, Michigan, in 2010, I worked as the head breakfast pastry chef for a busy pastry shop. When I got involved with Greenbush Brewing Company in 2013, I worked the hot line, cold prep, pastry chef, and was finally promoted to executive chef.

Q: What are some of your duties?

A: My main job is to run our food hub, which produces and distributes food to the three other restaurants affiliated with Greenbush. Other main tasks include meeting and coordination with my kitchen managers, vendor management and price negotiating, improving food and labor cost, employee relations, daily restaurant operations, menu creation, and quality control. I also plan and execute all of our on-site private events and off-site catering and festivals.

Q: Can you describe your typical workday?

A: Craziness, especially in the summer. On any given day, I could be doing any number of the tasks listed above. I don't get to go home for more than a few hours for three days straight sometimes. During the slower season, I am often at work by 8 a.m. and done by 5 or 6 p.m. I have most nights and two days off a week.

Q: What do you like most about your job?
A: My job is fun, exciting, and challenging. I have accomplished a lot and have the ability to accomplish a lot more. I work for a company that I am proud of because of the culture and brand I have helped create. The money is pretty good, too.

Q: What do you like least about your job?
A: The hours and time away from my family is the hardest part. It is very stressful, and it causes me to live a pretty unhealthy life style.

Q: What personal qualities do you find most valuable for this type of work?
A: Hard work, perseverance, and patience is absolutely required. The ability to stay calm and think quickly in stressful situations is a big plus. One also must have passion for the industry otherwise you will burn out quickly.

Q: What advice do you have for students who might be interested in this career?
A: It is fun and rewarding if you are willing to put in the effort. Don't be fooled by the glamorous-looking food TV program offers. The restaurant industry is nothing like that. All those people on TV put in a lot of time and effort to succeed; you will need to also. And don't go into massive debt over culinary school. You will never get out of it unless you are part of the lucky few.

Other Jobs in Travel and Hospitality

Amusement park manager
Asset manager
Banquet sales manager
Bed and breakfast facility
  manager
Casino manager
Catering director
Club manager
Concierge
Conference and event sales
  manager
Convention manager
Credit manager
Digital marketing manager
Director of room operations
Duty manager
Event coordinator
Floor supervisor
Food service manager
Front office supervisor

Gaming surveillance officer
Group sales manager
Guest services manager
Health care food service director
Human resource manager
Institutional food service
  manager
Meeting planner
Public relations manager
Purchasing director
Reservations supervisor
Restaurant manager
Revenue manager
Room service supervisor
Sales and marketing director
School food service director
Shore excursions manager
Tourism manager
Wedding sales manager
Welcome desk manager

Editor's note: The online *Occupational Outlook Handbook* of the US Department of Labor's Bureau of Labor Statistics is an excellent source of information on jobs in hundreds of career fields, including many of those listed here. The *Occupational Outlook Handbook* may be accessed online at www.bls.gov/ooh.

Index

Note: Boldface page numbers indicate illustrations.

About the Author

Stuart A. Kallen is the author of more than 350 nonfiction books for children and young adults. He has written on topics ranging from the theory of relativity to the art of electronic dance music. In addition, Kallen has written award-winning children's videos and television scripts. In his spare time he is a singer, songwriter, and guitarist in San Diego.